Copyright 2019 Gumdrop Press

All rights reserved.

ISBN-13: 978-1-945887-77-2

No part of this book may be reproduced in any written, electronic, or photocopied form without written permission of the publisher or author.

Every effort has been made to ensure the accuracy of the information contained in this book. The author and publisher disclaim liability to any party for any loss, damage, or disruption caused by errors or omissions that may result from the use of information contained within, whether such errors or omissions result from negligence, accident, or any other cause.

JANUARY 2020

Sun.	Mon.	Tue.	Wed.	Thu.	Fri.	Sat.
			1 New Year's Day	2	3	4
5	6	7	8	9	10	11
12	13	14	15	16	17	18
19	20 Martin Luther King, Jr. Day	21	22	23	24	25
26	27	28	29	30	31	

FEBRUARY 2020

Sun.	Mon.	Tue.	Wed.	Thu.	Fri.	Sat.
						1
2	3	4	5	6	7	8
9	10	11	12	13	14 Valentine's Day	15
16	17 Presidents' Day	18	19	20	21	22
23	24	25	26	27	28	29

MARCH 2020

Sun.	Mon.	Tue.	Wed.	Thu.	Fri.	Sat.
1	2	3	4	5	6	7
8 **Daylight Saving Time Begins**	9	10	11	12	13	14
15	16	17 **St. Patrick's Day**	18	19	20	21
22	23	24	25	26	27	28
29	30	31				

APRIL 2020

Sun.	Mon.	Tue.	Wed.	Thu.	Fri.	Sat.
			1	2	3	4
5	6	7	8	9	10	11
12 Easter	13	14	15	16	17	18
19	20	21	22	23	24	25
26	27	28	29	30		

MAY 2020

Sun.	Mon.	Tue.	Wed.	Thu.	Fri.	Sat.
					1	2
3	4	5	6	7	8	9
10 **Mother's Day**	11	12	13	14	15	16
17	18	19	20	21	22	23
24	25 **Memorial Day**	26	27	28	29	30
31						

JUNE 2020

Sun.	Mon.	Tue.	Wed.	Thu.	Fri.	Sat.
	1	2	3	4	5	6
7	8	9	10	11	12	13
14	15	16	17	18	19	20
21 **Father's Day**	22	23	24	25	26	27
28	29	30				

JULY 2020

Sun.	Mon.	Tue.	Wed.	Thu.	Fri.	Sat.
			1	2	3 **Independence Day Observed**	4 **Independence Day**
5	6	7	8	9	10	11
12	13	14	15	16	17	18
19	20	21	22	23	24	25
26	27	28	29	30	31	

AUGUST 2020

Sun.	Mon.	Tue.	Wed.	Thu.	Fri.	Sat.
						1
2	3	4	5	6	7	8
9	10	11	12	13	14	15
16	17	18	19	20	21	22
23	24	25	26	27	28	29
30	31					

SEPTEMBER 2020

Sun.	Mon.	Tue.	Wed.	Thu.	Fri.	Sat.
		1	2	3	4	5
6	7 Labor Day	8	9	10	11	12
13	14	15	16	17	18	19
20	21	22	23	24	25	26
27	28	29	30			

OCTOBER 2020

Sun.	Mon.	Tue.	Wed.	Thu.	Fri.	Sat.
				1	2	3
4	5	6	7	8	9	10
11	12 Columbus Day	13	14	15	16	17
18	19	20	21	22	23	24
25	26	27	28	29	30	31 Halloween

NOVEMBER 2020

Sun.	Mon.	Tue.	Wed.	Thu.	Fri.	Sat.
1 Daylight Saving Time Ends	2	3 Election Day	4	5	6	7
8	9	10	11 Veterans Day	12	13	14
15	16	17	18	19	20	21
22	23	24	25	26 Thanksgiving Day	27	28
29	30					

DECEMBER 2020

Sun.	Mon.	Tue.	Wed.	Thu.	Fri.	Sat.
		1	2	3	4	5
6	7	8	9	10	11	12
13	14	15	16	17	18	19
20	21	22	23	24	25 Christmas Day	26
27	28	29	30	31 New Year's Eve		

www.ingramcontent.com/pod-product-compliance
Lightning Source LLC
Chambersburg PA
CBHW080212040426
42333CB00043B/2621